Informing the legislative debate since 1914

Farm Commodity Provisions in the 2014 Farm Bill (P.L. 113-79)

Dennis A. Shields
Specialist in Agricultural Policy

March 28, 2014

Congressional Research Service

7-5700

www.crs.gov

R43448

Summary

The farm commodity program provisions in Title I of the Agricultural Act of 2014 (P.L. 113-79, the 2014 farm bill) include three types of support for crop years 2014-2018:

- **Price Loss Coverage (PLC) payments,** which are triggered when the national average farm price for a covered commodity (e.g., wheat, corn, soybeans, rice, and peanuts) is below its statutorily fixed "reference price";

- **Agriculture Risk Coverage (ARC) payments**, as an alternative to PLC, which are triggered when crop revenue is below its guaranteed level based on a multi-year moving average of historical crop revenue; and

- **Marketing Assistance Loans (MALs),** which offer interim financing for the loan commodities (covered crops plus several others) and, if prices fall below loan rates set in statute, additional low-price protection, sometimes paid as **loan deficiency payments (LDPs)**.

The enacted 2014 farm bill eliminated "direct payments," which were provided annually to producers and landowners of covered commodities from 1996 to 2013 based on historical production and a fixed payment rate set in statute. All farm program support now consists of variable payments. The PLC and ARC programs are enhanced relative to their predecessors via higher reference prices for PLC and more "local" coverage for ARC (whereby payments are triggered by county or individual losses rather than at the state level). In a major departure from previous farm bills and in response to a trade dispute with Brazil, upland cotton is no longer a covered commodity, with support for that crop now provided by a new crop insurance policy called the Stacked Income Protection Plan (STAX).

Approximately three-fourths of the 10-year, $47 billion in savings associated with the elimination of 2008 farm bill commodity programs was used to offset the costs of revising the overall farm safety net, specifically farm programs in Title I of the 2014 farm bill, adding permanent disaster assistance (also in Title I), enhancing the permanently authorized federal crop insurance program (Title XI), and enhancing the Noninsured Crop Disaster Assistance Program or NAP (Title XII). Crop insurance is available for more than 100 crops, including fruits and vegetables, and is designed primarily to cover losses from natural disasters. Farm programs do not require a participation fee, while crop insurance requires participating farmers to pay part of program costs.

The enacted 2014 farm bill sets a $125,000 per person cap on the total of PLC, ARC, marketing loan gains and loan deficiency payments. The limit applies to the total from all covered commodities except peanuts, which has a separate $125,000 limit. Also, to be eligible for payments, persons must be "actively engaged" in farming. The 2014 farm bill instructs USDA to write regulations (beginning with the 2015 crop year) that define "significant contribution of active personal management" to more clearly and objectively implement existing law. A single, total adjusted gross income (AGI) limit for payment eligibility is established at $900,000, which is less than the sum of the two separate limits (farm and non-farm) in the 2008 farm bill.

The Congressional Budget Office cost estimate (score) of the Title I provisions represents five-year savings of $6.3 billion and 10-year savings of $14.3 billion (both relative to baseline projections made in May 2013 assuming continuation of the 2008 farm bill). If these scores are added to the 2013 CBO baseline of budget outlays used to write the farm bill, then CBO's estimated cost of Title I is $23.6 billion for FY2014-2018 and $44.5 billion over 10 years.

Contents

Figures

Tables

Appendixes

Contacts

Introduction

On February 7, 2014, President Obama signed into law a new five-year omnibus farm bill, the Agricultural Act of 2014 (H.R. 2642; P.L. 113-79, the 2014 farm bill). The House had voted, 251-166, to approve the conference report (H.Rept. 113-333) on January 29, 2014, and the Senate approved the conference report on February 4, 2014, by a vote of 68-32. The U.S. Department of Agriculture (USDA) is now implementing the provisions, most of which take effect this year.

This report describes the farm commodity programs in Title I of the 2014 farm bill for "covered commodities" such as wheat, corn, soybeans, rice, and peanuts. Producer support is provided for the 2014-2018 crop years primarily through either statutory ("reference") prices or historical revenue guarantees based on the five most recent years of crop prices and yields.[1]

Important policy developments in the new law are also discussed and compared to prior law. The most significant policy change for commodity programs in the 2014 farm bill was the elimination of fixed direct payments and the enhancement of variable payments to farmers and landowners when crop prices or revenue declines.

Table A-1 provides detailed descriptions of farm commodity program provisions compared with prior law. For more on the legislative history of the 2014 farm bill and a side-by-side summary of crop insurance and all other farm bill provisions, see CRS Report R43076, *The 2014 Farm Bill (P.L. 113-79): Summary and Side-by-Side.*

Background

Policy Rationale for Farm Subsidies

Federal farm support began in the 1930s through Depression-era efforts to raise farm household income when commodity prices were low because of prolonged weak consumer demand. While initially intended to be a temporary effort, the commodity support programs survived, but have been modified away from supply control and management of commodity stocks (which was designed to prop up prices) into direct income support payments. The 2014 farm bill continues traditional farm support via variable payments relative to statutory price levels or historical crop revenue.

Proponents of farm commodity programs argue that federal involvement in the sector is needed to stabilize and support farm incomes by shifting some of the risks to the federal government. These risks include short-term market price instability and longer-term capacity adjustments. Proponents see the goal of farm policy as maintaining the economic health of the nation's farm sector so that it can use its comparative advantage in feeding the nation and competing in the global market for food and fiber. Critics argue that farm commodity programs waste taxpayer dollars, distort production of certain crops, capitalize benefits to the owners of the resources,

---

[1] Separately, under the federal crop insurance program, covered commodities and more than 100 other crops, including fruits and vegetables, are also eligible for subsidized crop insurance, which provides within-year yield (or revenue) protection. That is, the guarantee is set each year just prior to planting the crop.

encourage concentration of production, and comparatively harm smaller domestic producers and farmers in lower-income foreign nations.

Authorizing Legislation

The authority for USDA to operate farm commodity programs comes from three permanent laws, as amended: the Agricultural Adjustment Act of 1938 (P.L. 75-430), the Agricultural Act of 1949 (P.L. 81-439), and the Commodity Credit Corporation (CCC) Charter Act of 1948 (P.L. 80-806). Congress typically alters these laws through multi-year omnibus farm bills to address current market conditions, budget constraints, or other concerns.

If a new farm bill is not enacted when an old one expires, farm programs would revert to the permanent laws mentioned above for most of the major program crops. Under permanent law, eligible commodities would be supported at levels much higher than they are now, and many of the currently supported commodities might not be eligible. Since reverting to permanent law is incompatible with current national economic objectives, global trading rules, and federal budgetary policies, pressure builds at the end of one farm bill to enact another.[2]

The 2014 farm bill (P.L. 113-79) contains the most recent version of the farm commodity support programs. It supersedes the commodity provisions of previous farm bills, and suspends the relevant price support provisions of permanent law.

Eligible Commodities

Federal support exists for about two dozen farm commodities representing about one-third of gross farm sales. During FY2005-FY2014, five crops (corn, cotton, wheat, rice, and soybeans) accounted for about 90% of these payments.[3]

- Under the 2014 farm bill, the "covered commodities" are the primary crops eligible for farm support: wheat, oats, and barley (including wheat, oats, and barley used for haying and grazing); corn, grain sorghum, long grain rice, medium grain rice, and pulse crops (dry peas, lentils, small chickpeas, and large chickpeas); soybeans, other oilseeds (including sunflower seed, rapeseed, canola, safflower, flaxseed, mustard seed, crambe, and sesame seed), and peanuts.

  In a major departure from all previous farm bills and in response to a trade dispute with Brazil, upland cotton is no longer a covered crop, with support for that crop now provided by a new crop insurance policy called the Stacked Income Protection Plan (STAX).[4]

- "Loan commodities" include all of the "covered commodities" plus upland cotton, extra long staple cotton, wool, mohair, and honey. These commodities are eligible for the marketing loan program only.

---

[2] For more background on the consequences of reverting to permanent law, see CRS Report R42442, *Expiration and Extension of the 2008 Farm Bill.*

[3] U.S. Department of Agriculture, Farm Service Agency, *CCC Budget Essentials*, FY2014 CCC Table 35, http://www.fsa.usda.gov/FSA/webapp?area=about&subject=landing&topic=bap-bu-cc.

[4] For background, see CRS Report R43336, *Status of the WTO Brazil-U.S. Cotton Case.*

- The 2014 farm bill replaces the dairy product price support program and Milk Income Loss Contract (MILC) payments with new dairy programs to (1) protect producer margins (milk prices minus feed costs), and (2) buy excess dairy products to boost demand when margins drop below certain levels.

- Sugar support is indirect through import quotas, price guarantees, and domestic marketing allotments. No direct payments are made to growers and processors. There was no change to the sugar program in the 2014 farm bill. See CRS Report R42551, *Sugar Provisions of the 2014 Farm Bill (P.L. 113-79)*.

Meats, poultry, fruits, vegetables, nuts, hay, and nursery products (about two-thirds of farm sales) do not receive direct support or payments under the commodity programs of the farm bill. However, livestock and tree fruit producers receive disaster support under Title I of the 2014 farm bill. (See **Table A-1** and CRS Report RS21212, *Agricultural Disaster Assistance*, for a description of disaster programs.) Also, under the permanently authorized federal crop insurance program, subsidized crop insurance is available for more than 100 crops, including fruits and vegetables which are not supported by farm programs. Crop insurance is designed primarily to cover losses from natural disasters and within-season price or revenue declines (see CRS Report R40532, *Federal Crop Insurance: Background*).

Definition of "Farm"

The definition of "farm" used to administer the commodity programs is different from other statistical or perceived definitions of farms. Under Farm Service Agency (FSA) regulations, a "farm" for program payment purposes is one or more tracts of land considered to be a separate operation.[5] Land in a farm does not need to be contiguous; however, all tracts within a farm must have the same operator and the same owner (unless all owners agree to combine multiple tracts into a single FSA farm). Thus, one producer may be operating several "farms" if he/she is renting land from several landlords, or has purchased land in several tracts.

Base Acres

For the purpose of calculating program payments, the term "base acres" is the historical planted acreage on each FSA farm, using a multi-year average from as far back as the 1980s.[6] Technically, a farm's base with respect to a covered commodity is the number of acres in effect under the 2008 farm bill (7 U.S.C. 8702, 8751) as of September 30, 2013, subject to any reallocation, adjustment, or reduction under the 2014 farm bill. Base is calculated for each covered commodity and transfers to the new owner when land is sold, making the new landowner eligible for farm programs.

Because a farmer's actual plantings may differ from farm base acres, program payments may not necessarily align with financial losses associated with market prices or crop revenue. In order to better match program payments with farm risk, the 2014 farm bill provides farmers with a one-time opportunity to update individual crop base acres by reallocating acreage within their current

---

[5] 7 C.F.R. 718.2.

[6] Base acre provisions since 1981 are described in Edwin Young et al., *Economic Analysis of Base Acre and Payment Yield Designations Under the 2002 U.S. Farm Act*, USDA Economic Research Service, September 2005, pp. 36-41, http://www.ers.usda.gov/publications/err-economic-research-report/err12.aspx#.UzL3jYUq4Vc.

base to match their actual crop mix (plantings) during 2009-2012. Farmers can also choose to not reallocate their base if they expect payments to be maximized under their current base. In the case of cotton, which is no longer a covered commodity, former cotton base acres are renamed "generic base" and added to a producer's base for potential payments if a covered crop is planted on the farm.[7]

"Partially Decoupled" Payments

Payments under the new programs in the 2014 farm bill are made on base acres, not current plantings.[8] This feature—decoupling payments from current plantings—is intended to better comply with World Trade Organization (WTO) rules on domestic support and to minimize any influence on producer behavior and prevent any subsequent market distortion. The payments are considered "partially decoupled" because the payment amount remains connected to current market prices. In the 2008 farm bill, farm payments were calculated using either base or planted area, depending upon the program.

Eligible Producers

The 2014 farm bill defines a producer (for purposes of farm program benefits) as an owner-operator, landlord, tenant, or sharecropper that shares in the risk of producing a crop and is entitled to a share of the crop produced on the farm. For payment eligibility, a term commonly used in federal regulations is "actively engaged in farming," which generally means providing significant contributions of capital (land or equipment) and labor and/or management, and receiving a share of the crop as compensation. The 2014 farm bill requires USDA to write new regulations that define "significant contribution of active personal management." See "Payment Limits," below.

Producers do not pay to participate in farm programs. However, an individual must comply with certain conservation and planting flexibility rules. Conservation rules include protecting wetlands, preventing erosion, and controlling weeds. Planting flexibility rules allow crops other than the program crop to be grown, but under the 2014 farm bill, eligible payment acreage is reduced when fruits, vegetables, or wild rice are planted in excess of 15% of base acres (or 35% depending upon a farmer's program choice discussed below). Also, a producer on a farm may not receive farm program payments if the sum of the base acres on the farm is 10 acres or less.

A farm enterprise usually involves some combination of owned and rented land. Two types of rental arrangements are common: cash rent and share rent. Under cash rental contracts, the tenant pays a fixed cash rent to the landlord. The landlord receives the same rent, bears no risk in production, and thus is not eligible to receive program payments. The tenant bears all of the risk, takes all of the harvest, and receives all of the government subsidy.

---

[7] Specifically, for each crop year, generic base acres are attributed to (i.e. temporarily designated as) base acres to a particular covered commodity base in proportion to that covered crop's share of total plantings of all covered commodities in that year. However, if the total number of acres planted to all covered commodities on the farm does not exceed the generic base acres on the farm, only the amount of acreage actually planted to a covered commodity is eligible for payment.

[8] The exception is payments associated with generic base acres, whereby current plantings can affect payment acreage.

Under share rental contracts, the tenant usually supplies most or all of the labor and machinery, while the landlord supplies land and perhaps some machinery or management. Both the landlord and the tenant bear risk in producing a crop and receive a portion of the harvest.[9] Both are eligible to share in the government subsidy.

Even though tenants might receive all of the government payments under cash rent arrangements, they might not keep all of the benefits if landlords demand higher rent. Economists widely agree that a large portion of government farm payments passes through to landlords, since government payments boost the rental value of land. The amount of total land in farms rented by farm operators has ranged between 34% and 43% of farmland during 1964-2007.[10]

Eliminated 2008 Farm Bill Programs

Under the enacted 2014 farm bill (P.L. 113-79), farm support for traditional program crops is restructured by eliminating the direct payment (DP) and counter-cyclical payment (CCP) programs, and the Average Crop Revenue Election (ACRE) program. For the 1996 through 2013 crop years, direct payments were made to producers and landowners based on historical production of corn, wheat, soybeans, cotton, rice, peanuts, and other "covered" crops. Direct payments lost political support in recent years because recipients did not need to suffer an income loss in order to receive a payment. Approximately three-fourths of the 10-year, $47 billion in savings associated with the elimination of current farm programs was used to offset the costs of revising farm programs in Title I of the 2014 farm bill, adding permanent disaster assistance (also in Title I), enhancing the permanently-authorized federal crop insurance program (Title XI), and enhancing the Noninsured Crop Disaster Assistance Program or NAP (Title XII).

Farm Commodity Program Provisions

The farm commodity program provisions in Title I of the 2014 farm bill include three types of support for crop years 2014-2018:

- **Price Loss Coverage (PLC) payments**, which are triggered when the national average farm price for a covered commodity is below its statutorily-fixed "reference price";

- **Agriculture Risk Coverage (ARC) payments**, as an alternative to PLC, which are triggered when crop revenue is below its guaranteed level based on a multi-year moving average of historical crop revenue; and

- **Marketing Assistance Loans (MALs)** that offer interim financing for the loan commodities (covered crops plus several others as indicated above) and, if prices fall below loan rates set in statute, additional low-price protection, sometimes paid as **loan deficiency payments (LDPs)**.

---

[9] For example, a typical share rental arrangement in some regions is a two-thirds/one-third split of the crop harvested, with the landlord supplying all of the land and one-third of the cost of certain inputs such as fertilizer. The tenant supplies all of the labor and pays the remaining share of the input costs. Management decisions, such as crop diversification, are usually made jointly.

[10] C. Nickerson et al., "Trends in U.S. Farmland Values and Ownership," Economic Information Bulletin Number 92, USDA Economic Research Service, February 2012, http://www.ers.usda.gov/media/377487/eib92_2_.pdf.

Farmers with base acres of covered commodities have a one-time irrevocable decision to choose between PLC and "county" ARC (based on a county guarantee) on a commodity-by-commodity basis for each farm. Alternatively, all covered crops on a farm can be enrolled in "individual" ARC, which is based on a farm-level guarantee. (See "Agriculture Risk Coverage (ARC)," below.) If no choice is made, the producer forfeits any payments for the 2014 crop year and the farm is enrolled automatically in PLC for the 2015-2018 crop years. The "optimal" decision depends in part on expected prices through 2018 relative to guarantees in each program.

The PLC and ARC programs are similar conceptually to the 2008 farm bill's counter-cyclical payment (CCP) program and Average Crop Revenue Election (ACRE) program, respectively. However, compared with the previous programs, they have enhanced levels of protection from low prices (i.e., higher price parameters in PLC) or revenue loss (i.e., county- or farm-level guarantees for ARC rather than state-level in ACRE).

PLC and ARC payments are proportional to base acres, and not planted acres.[11] Payments are made with a lag of approximately one year as annual price and yield data are compiled for USDA's calculations. USDA is to issue payments beginning October 1 after the end of each marketing year, which varies by crop. For example, the marketing year for corn harvested in fall of 2014 ends in August 2015.

Marketing assistance loans are available for covered crops and other loan commodities. The program continues mostly unchanged from the 2008 farm bill, with loan rates set at relatively low levels compared to historical prices.

All three types of payments are subject to a combined payment limit of $125,000 per person. Also, the income limit for program eligibility is $900,000 for adjusted gross income (three-year average). See "Payment Limits" and "Adjusted Gross Income (AGI) Limit," below.

Price Loss Coverage (PLC)

For each covered commodity on a farm, producers may select the Price Loss Coverage (PLC) program to receive a payment on 85% of base acres when the annual national average farm price is below the reference price set in statute. This option could be attractive if farmers expect farm prices to drop below statutory minimums.

Payments are proportional to a farm's base acres, historical farm yield, and the difference between the reference price and the annual farm price. Hence payments are generally "decoupled" from planted acreage and actual yield but not price. PLC payments operate the same as CCPs under the 2008 farm bill, which have been reported to the WTO by the United States as "amber box" subsidies, and thus limited in size together with other amber box subsidies.

Commodity groups successfully argued for an increase in reference prices relative to the payment trigger levels in the 2008 farm bill (i.e., target price minus direct payment rate). For example, the payment trigger level has been raised by 51% for wheat, 57% for corn, 51% for soybeans, 72%

---

[11] The exception is payments on "generic" base acres (formerly cotton base acres), which are directly attributable to the planted crop(s) until the total of the covered commodities planted on the farm exceed the generic base. If covered crop plantings are greater than the generic base, payment acres are attributed based on the proportion of the covered commodities planted.

for rice (98% for temperate Japonica rice), and 17% for peanuts. Reference prices and a comparison with 2008 farm bill parameters for each covered commodity are shown in **Table 1**.

The PLC payment formula is 85% *times* the number of base acres *times* historical payment yield *times* the difference between the reference price and the annual farm price (or loan rate if higher). See **Figure 1** for a graphical interpretation of the formula and **Figure 2** for a hypothetical example for rice. The historical payment yield is equal to 90% of the 2008-2012 average yield per planted acre for the farm. As an alternative, the producer can keep the program yield used for calculating CCPs in the 2008 farm bill (generally based on 1998-2001 yields).

Table 1. Loan Rates and Reference Prices in the 2014 Farm Bill

| | 2008 and 2014 farm bills | Price at which a payment is triggered: | | |
| | | 2008 farm bill | 2014 farm bill | |
	Loan Rate	**Target Price minus Direct Payment Rate**	**Reference Price**	**% change from 2008 farm bill**
Wheat, $/bu	2.94	4.17 – 0.52 = **3.65**	5.50	+51%
Corn, $/bu	1.95	2.63 – 0.28 = **2.35**	3.70	+57%
Sorghum, $/bu	1.95	2.63 – 0.35 = **2.28**	3.95	+73%
Barley, $/bu	1.95	2.63 – 0.24 = **2.39**	4.95	+107%
Oats, $/bu	1.39	1.79 – 0.024 = **1.766**	2.40	+36%
Upland Cotton, $/lb	2008 farm bill: 0.52 2014 farm bill: 0.45 to 0.52	0.7125 – 0.0667 = **0.6458**	n.a.	n.a.
ELS cotton, $/lb	0.7977	n.a.	n.a.	n.a.
Rice, $/cwt	6.50	10.50 – 2.35 = **8.15**	14.00; 16.10 for temperate japonica	+72%; +98% for temperate japonica
Soybeans, $/bu	5.00	6 – 0.44 = **5.56**	8.40	+51%
Minor oilseeds, $/lb	0.1009	0.1268 – 0.008 = **0.1188**	0.2015	+70%
Peanuts, $/ton	355	495-36 = **459**	535	+17%
Peas, dry, $/cwt	5.40	8.32 – 0 = **8.32**	11.00	+32%
Lentils, $/cwt	11.28	12.81 – 0 = **12.81**	19.97	+56%
Sm.chickpeas, $/cwt	7.43	10.36 – 0 = **10.36**	19.04	+84%
Lg.chickpeas, $/cwt	11.28	12.81 – 0 = **12.81**	21.54	+68%
Wool, graded, $/lb	1.15	n.a.	n.a.	n.a.
Wool, nongraded	0.40	n.a.	n.a.	n.a.
Mohair $/lb	4.20	n.a.	n.a.	n.a.
Honey, $/lb	0.69	n.a.	n.a.	n.a.
Sugar, raw cane, $/lb	0.1875	n.a.	n.a.	n.a.
Sugar, beet, $/lb	0.2409	n.a.	n.a.	n.a.

Source: CRS.

Note: n.a. = not applicable.

Figure 1. Price Loss Coverage (PLC)

(makes payment when national average farm price drops below the reference price)

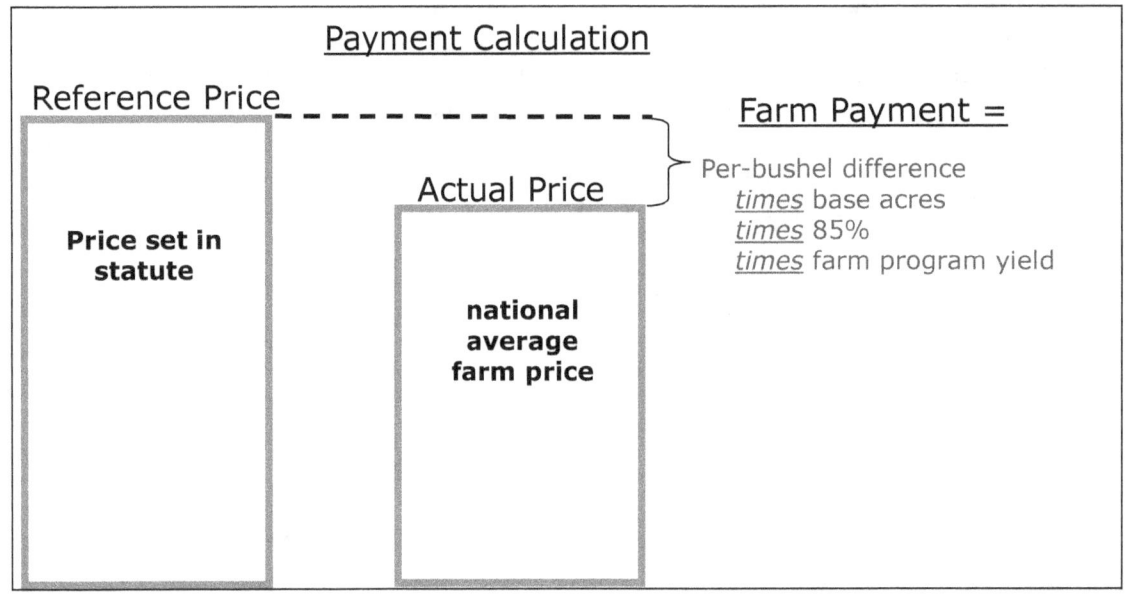

Source: CRS.

Note: In a declining market, the per-bushel payment rate increases until the farm price drops below the loan rate. At this point, benefits under the Marketing Assistance Loan Program may become available.

Figure 2. Price Loss Coverage (PLC): Low Price Scenario for Rice

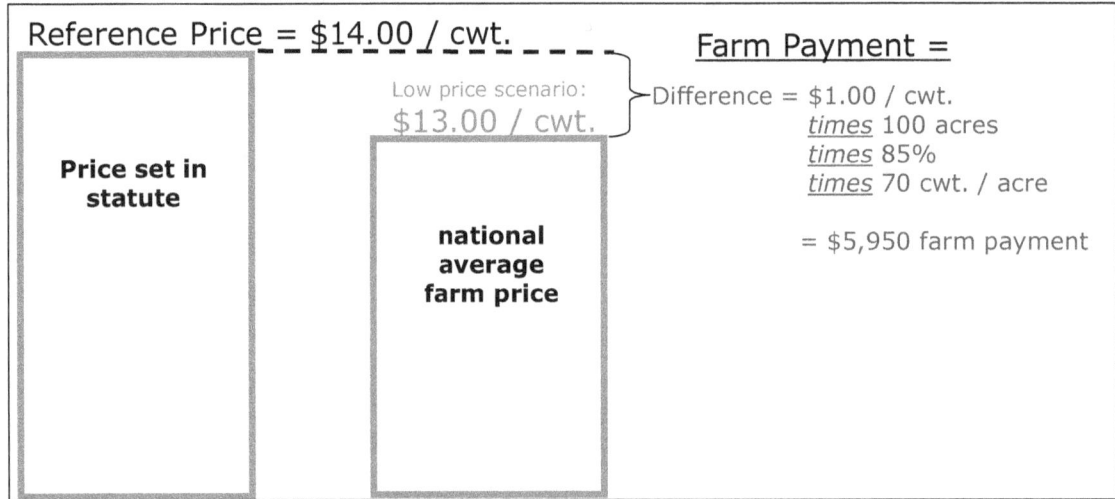

Source: CRS.

Notes: In a declining market, the per-bushel payment rate increases until the farm price drops below the loan rate ($6.50/cwt. for rice). If market prices decline further, benefits under the Marketing Assistance Loan Program may become available.

Agriculture Risk Coverage (ARC)

Producers more concerned about declines in crop revenue (i.e., yield times price) than just price can select the county Agriculture Risk Coverage (ARC) program as an alternative to PLC for each covered commodity. Payments are made on 85% of base acres when annual crop revenue is less than 86% of its historical level.

If farmers prefer individual farm level protection, they must enroll *all* covered crops on the farm in the ARC-individual coverage option instead of selecting between PLC and county ARC for each crop.

County ARC

For producers choosing between ARC and PLC on each covered commodity on a farm, the county ARC program has a county revenue guarantee, and only a crop revenue loss at the county level triggers a payment. For ARC county coverage, payments are made on 85% of base acres when actual county crop revenue drops below the county revenue guarantee, which is 86% of historical or "benchmark" revenue. The benchmark revenue per acre is equal to the average historical county yield for the most recent 5 crop years (excluding the years with the highest and lowest yields, or "Olympic average") times the national average market price received by producers during the 12-month marketing year for the most recent 5 crop years (excluding the years with the highest and lowest prices). With the guarantee set at 86%, the producer absorbs the first 14% of the shortfall, and the government absorbs the next 10% of revenue shortfall. (The per-acre payment rate is capped at 10% of benchmark revenue.) Remaining losses are backstopped by crop insurance if purchased at sufficient coverage levels by the producer and by the marketing assistance loan program.

The county ARC payment formula is 85% *times* the number of base acres *times* the difference between the county revenue guarantee and the actual crop revenue. See **Figure 3** for a graphical interpretation of the formula and **Figure 4** for a hypothetical example for corn.

Individual ARC

Farm level protection is provided if producers enroll *all* covered crops on the farm in the ARC-individual coverage option, which uses individual farm yields for each covered crop (which are more variable than county averages) and aggregates all crop revenue into a single, whole-farm guarantee. Individual coverage was not available for ACRE in the 2008 farm bill; farm-level coverage was provided instead by the Supplemental Revenue Assistance (SURE) disaster program (not reauthorized under the 2014 farm bill).

The individual ARC payment formula is 65% *times* the number of total base acres for the farm *times* the difference between the revenue guarantee and the actual crop revenue. The calculation for the guarantee and actual revenue are based on the aggregation of all covered crops on the farm using individual farm yields instead of county yields.[12]

---

[12] For a description of the calculations, see 2014 farm bill Section 1117(b)(2) and Section 1117(c)(3) in the **Appendix**. An example of ARC-individual coverage is available in Jonathan Coppess, "Farm Bill Programs in the 2014 Farm Bill," Farmdoc Webinar, March 5, 2014, http://www.farmdoc.illinois.edu/webinars.

Figure 3. Agriculture Risk Coverage (ARC)–County Coverage

(payment when actual county-wide revenue drops below 86% of historical revenue ["shallow loss"])

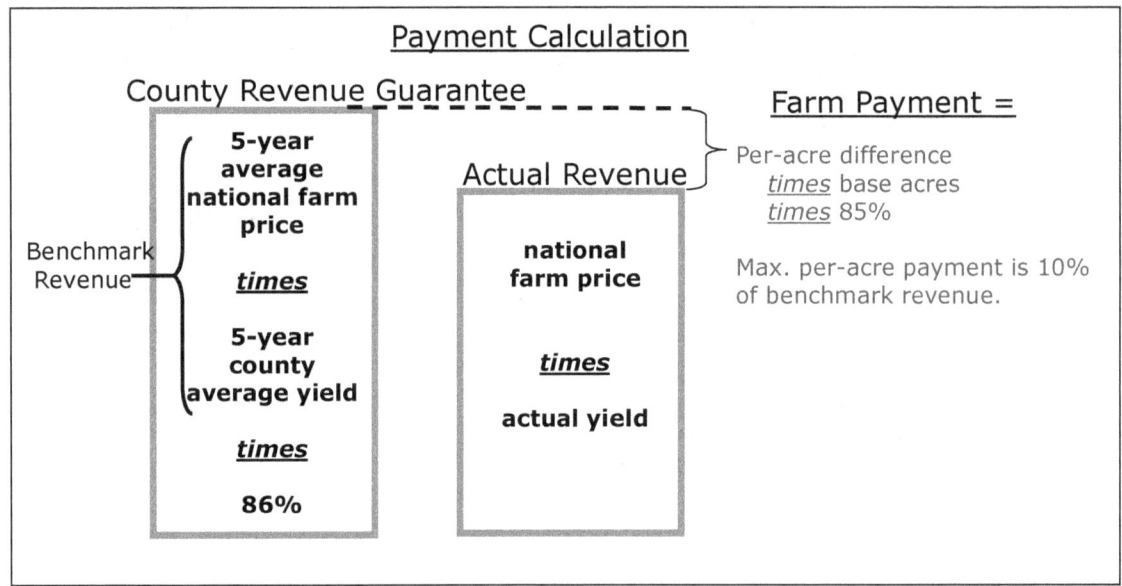

Source: CRS.

Notes: Five-year averages exclude high and low years. Instead of an ARC county guarantee on a crop-by-crop basis, farmers can select a farm-level guarantee for all covered crops on a farm. Payment acreage is reduced to 65% of base acres, and a single, whole farm guarantee (and payment) is calculated as a weighted average for all crops (i.e., not on a crop-by-crop basis).

Figure 4. Agriculture Risk Coverage (ARC)–Low Revenue Scenario for Corn

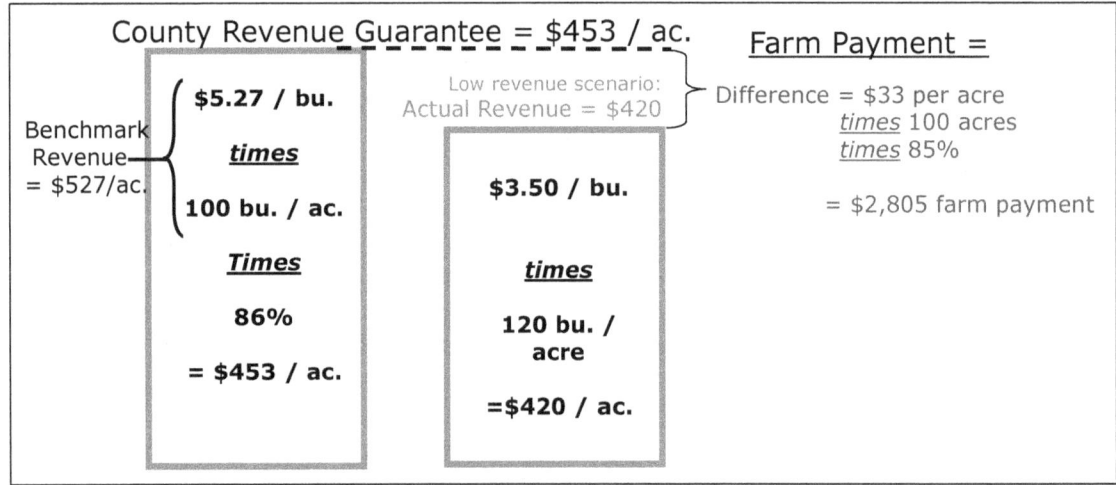

Source: CRS.

Notes: Assumes five-year average price (excluding high and low years) is $5.27 per bushel and five-year average yield (excluding high and low years) is 100 bushels per acre.

Marketing Assistance Loan Program

The Marketing Assistance Loan (MAL) program provides additional financial benefits to farmers in the form of a guaranteed floor price for qualifying field crops, in addition to providing short-term financing. The process begins with a government loan to participating farmers of designated crops (covered commodities, plus upland cotton, extra long staple cotton, wool, mohair, and honey). The loan is made at a specified "per-unit" loan rate using the crop as collateral. This loan rate, in effect, establishes a price guarantee. Prior to loan maturity, if the local market price (called the "posted price") is at or above the loan rate, the farmer repays the loan principal and interest.[13] In contrast, when the posted price is below the loan rate, the farmer may repay the loan at that price (called the "loan repayment rate") and pocket the difference as a "marketing loan gain."[14] Or, rather than taking the loan when the posted price is below the loan rate, farmers may request a "loan deficiency payment," with the payment rate equal to the difference between the loan rate and the loan repayment rate.

Program benefits are available on the entire crop produced, which means a farmer receives no benefits in the event of a crop loss. This is in contrast to the other two programs (PLC and ARC) that make payments on historical acres and yields and therefore are not dependent on current production.

In the 2014 farm bill, for 2014-2018 crop years, loan rates remain the same as prior law except for upland cotton (see **Table 1** for loan rates). The loan rate for upland cotton is changed from $0.52 per lb. to the simple average of the adjusted prevailing world price for the two immediately preceding marketing years, but not less than $0.45 per pound or more than $0.52 per pound.

Given recent relatively high price levels, the MAL program has paid only limited benefits in recent years for most crops. As a result, some farmers have criticized loan rates as being too low relative to prevailing market prices. MAL program benefits, combined with payments under PLC and ARC, are subject to a payment limit of $125,000 per person for all covered commodities (except peanuts, which has a separate limit of $125,000). Benefits derived from loan forfeitures are exempt from the limit. The 2008 farm bill did not have a payment limit for MAL.

Cotton Not Eligible for Either PLC or ARC

Beginning with the 2014 farm bill, cotton is no longer a covered commodity and not eligible for PLC/ARC payments. Instead it is eligible for a new crop insurance policy called Stacked Income Protection or STAX. Cotton remains eligible for MAL but the loan rate was altered slightly as specified above.

The policy revision was sought by U.S. cotton producers in an attempt to resolve a long-running trade dispute with Brazil that requires changing the U.S. cotton support program so it does not

---

[13] The market price is the adjusted world market price for upland cotton and rice, the national posted price for peanuts, national or regional posted prices for pulse crops, and the posted county price for most other commodities.

[14] Farmers may also forfeit the crop pledged as collateral to the government at the end of the loan period. This type of loan is called nonrecourse. A few crops are eligible only for recourse loans (i.e., must be repaid at principal plus interest), including ELS cotton, seed cotton, and high-moisture grains. Recourse loans are not eligible for a subsidy but do offer low-interest financing.

distort international markets.[15] As part of the transition, farm payments are made for upland cotton for the 2014 crop year, and for 2015 if STAX is not available. Payment acres in 2014 equal 60% of 2013 cotton base acres and 36.5% of 2013 cotton base acres in 2015.

Separately, the 2014 farm bill specifies that upon resolution of the trade dispute, funds paid by the U.S. government to Brazil (as part of an agreement made in 2010) may be used for research conducted collaboratively between Brazil and USDA research agencies or with a college, university, or research foundation located in the United States. Among several provisions, the agreement required annual payments of $147.3 million from the United States (via the Commodity Credit Corporation, CCC) to Brazil in order to provide technical assistance and capacity-building for Brazil's cotton sector, but it explicitly excluded funding research.

Planting Fruits and Vegetables on Base Acres

Any crop may be planted without effect on base acres. However, payment acres on a farm are reduced in any crop year in which fruits, vegetables (other than mung beans and pulse crops), or wild rice have been planted on more than 15% of base acres (or 35% in the case of the individual coverage option for ARC). The reduction to payment acres is one-for-one for every acre in excess of these percentages. This allows a limited amount of fruits and vegetables without penalty. Unlike in the past when the reduction in payment acres began at the first acre of fruits and vegetables on base acres, the 2014 farm bill allows 15% (or 35% for individual ARC) of base acres to be planted to fruits and vegetables before the reduction in payment acres begins.

Payment Limits

The enacted 2014 farm bill sets a $125,000 per person cap on the total of PLC, ARC, marketing loan gains and loan deficiency payments. The limit applies to the total from all covered commodities except peanuts, with a separate $125,000 limit for peanuts (for both, limits are doubled with a spouse). This is in contrast to the 2008 farm bill, which had applied limits for each program, specifically $40,000 per person for direct payments, $65,000 for counter-cyclical and ACRE payments, and no limit on marketing loan gain or LDPs.

"Actively Engaged"

To be eligible for payments, persons must be "actively engaged" in farming. Actively engaged, in general, is defined as making a significant contribution of (i) capital, equipment or land, and (ii) personal labor or active personal management. Also, profits are to be commensurate with the level of contributions, and contributions must be at risk. Legal entities can be actively engaged if members collectively contribute personal labor or active personal management. Special classes allow landowners to be considered actively engaged if they receive income based on the farm's operating results, without providing labor or management. Under the 2008 farm bill, spouses were considered actively engaged if the other spouse meets the qualification, allowing payment limits to be doubled.

---

[15] See CRS Report R43336, *Status of the WTO Brazil-U.S. Cotton Case.*

The 2014 farm bill instructs USDA to write regulations that define "significant contribution of active personal management" to more clearly and objectively implement existing law. The regulation is to apply beginning with the 2015 crop year, and entities made solely of family members are exempt. This final provision differs from earlier Senate-passed and House-passed versions of the 2014 farm bill, which would have deleted "active personal management" and effectively required personal labor in the farming operation. The final 2014 farm bill provision instructs USDA to consider different limits for varying types of farming operations, based on considerations of size, nature, and management requirements of different farming types, changes in the nature of active personal management due to advancements in farming practices, and the impact of this regulation on the long-term viability of farming operations.

Adjusted Gross Income (AGI) Limit

To qualify for any commodity program benefits, recipients must pass an eligibility requirement based on adjusted gross income (AGI) used for federal taxes. The enacted 2014 farm bill establishes the AGI limit as a single, total AGI limit of $900,000 (using a three-year average). In contrast, the 2008 farm bill had two separate limits—farm and non-farm. The non-farm AGI limit was $500,000 to qualify for and receive any farm commodity program benefits, Milk Income Loss Contract (MILC) program, noninsured crop assistance (NAP), or disaster payments. The second limit was $750,000 on farm AGI to receive direct payments.

Some individuals who previously qualified for farm program payments might no longer qualify due to the lower overall limit ($900,000 compared with the combined 2008 farm bill limits of $500,000 and $750,000). However, others might regain eligibility if nonfarm income is high (i.e., between the previous non-farm limit of $500,000 and the new total limit of $900,000) and farm income is low enough to prevent total AGI from exceeding the 2014 farm bill AGI limit of $900,000.

Interaction with Federal Crop Insurance

Federal crop insurance intersects with farm programs when producers choose between the Agriculture Risk Coverage (ARC) and the Price Loss Coverage (PLC) programs. The ARC program is a "shallow loss" program that makes a payment when actual crop revenue is more than 14% below the ARC guarantee. For producers who select the PLC, "shallow loss" coverage is available by purchasing a new crop insurance product called Supplemental Coverage Option (SCO) authorized in Title XI of the 2014 farm bill. SCO is designed to cover part of the deductible on a producer's underlying crop insurance policy. SCO is not available for those enrolled in ARC.

Estimated Cost of the Commodity Title

Because spending on the farm commodity programs involves market-driven counter-cyclical payments, annual outlays can be highly variable. **Figure 5** shows that, from 1990 to 2013, commodity program outlays (including dairy programs and disaster payments) were about $5 billion both at the beginning and the end of this period. The high was $27 billion in 2000 when Congress authorized a large sum of "market loss assistance" payments in response to significant price declines in the crop sector. The average over the entire period was $10.1 billion per year.

In January 2014, the Congressional Budget Office (CBO) estimated that for the 2014-2023 period, farm commodity (and disaster program) spending will be $4.4 billion annually, well below the historical averages due to the elimination of direct payments and relatively strong market prices projected by CBO. Given the counter-cyclical design of farm programs, if commodity prices are higher than projected, government outlays will be below projected levels (and vice versa).

Figure 5. Outlays for Farm Commodity Program and Disaster Assistance

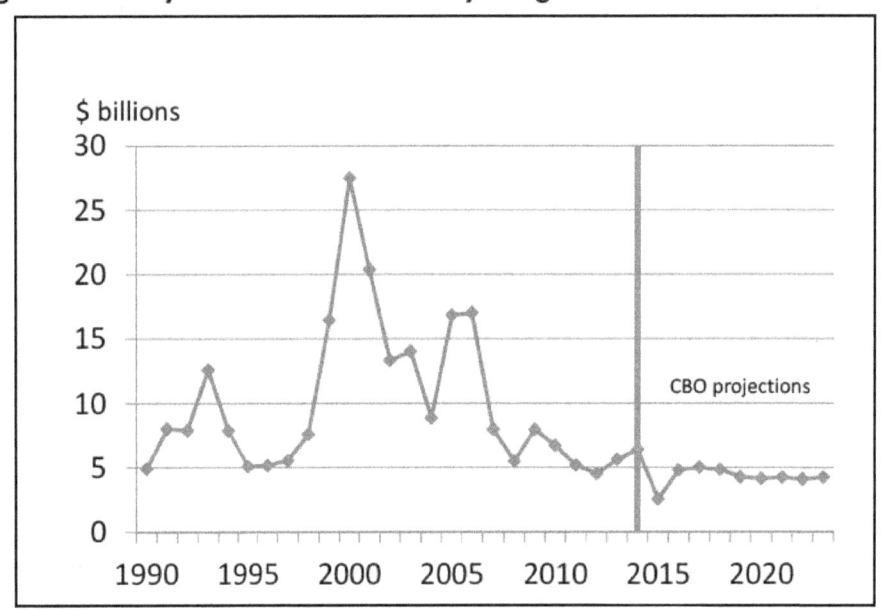

Source: CRS, using USDA (historical data) and CBO data (projections).

Notes: 2013 and prior years are actual outlays from USDA "Table 35, CCC Net Outlays by Commodity and Function," http://www.fsa.usda.gov/Internet/FSA_File/pb14_table_35x.xls., summation of rows for covered commodities and loan commodities, dairy, and disaster assistance; 2014-2023 are estimates based on the CBO May 2013 baseline projections for Title I plus CBO 2014 farm bill score for Title I, January 28, 2014.

Compared to the baseline of continuing the provisions of the 2008 farm bill, the CBO cost estimate (score) of the new provisions in Title I of the farm bill represents five-year savings of $6.3 billion and 10-year savings of $14.3 billion (both relative to baseline projections made in May 2013). If the scores of these changes are added to the 2013 baseline of budget outlays used to write the farm bill, then CBO's expected cost of Title I is $23.6 billion for FY2014-2018 and $44.5 billion over 10 years (**Table 2**). This includes the program crop commodities, dairy, disaster programs, and sugar (no program change).

The 5- and 10-year savings that are scored for all of Title I are the net result of various provisions that either score savings or cost more than prior law. The savings is the result of the elimination of direct, counter-cyclical, and ACRE payments. Also, payments for PLC/ARC are made with a one-year lag, which shifts about $3 billion of payments into a later fiscal year. This achieves savings in the 10-year budget window but does not reduce the total amount eventually paid to farmers. Some of the Title I savings are offset with costs of the new PLC and ARC programs, transition payments to producers of upland cotton (not eligible for PLC/ARC), reauthorized disaster programs, and a revised dairy program.

Table 2. Cost of Provisions in the Commodity Title of the 2014 Farm Bill

(in millions of dollars)

Description	5 years: FY2014-FY2018	10 years: FY2014-FY2023
CBO baseline, May 2013, Title I	**29,888**	**58,765**
CBO score of changes in Title I of the 2014 farm bill		
Provisions with net savings:		
Repeal direct payments	-18,153	-40,845
Repeal counter-cyclical payments	-489	-1,519
Repeal Average Crop Revenue Election payments	-2,494	-4,718
Provisions with net additional costs:		
Price Loss Coverage	5,115	13,124
Agriculture Risk Coverage	6,528	14,108
Transition payments to producers of upland cotton	558	558
Nonrecourse Marketing Assistance Loans	23	48
Dairy program	241	912
Supplemental Agriculture Disaster Assistance	2,166	3,674
Implementation	120	120
Loan implementation	54	230
Subtotal: CBO Score of 2014 Farm Bill changes, Title I	**-6,332**	**-14,307**
CBO Estimate of Total Cost of Title I	**23,556**	**44,458**

Source: CBO baseline and score of the conference agreement of H.R. 2642, the Agricultural Act of 2014, as reported on January 27, 2014.

Implementation

The 2014 farm bill provides $100 million to carry out Title I. A portion of the total ($3 million) is marked for educating producers on new commodity, dairy, and disaster programs; and $3 million for web-based program decision tools to help producers make their farm program choices (base reallocations, yield updates, PLC vs. ARC). Congress emphasized the importance of the Acreage Crop Reporting and Streamlining Initiative by making available an additional $10 million if USDA makes sufficient progress by September 2014 and another $10 million if it is fully implemented by September 30, 2015. USDA and university extension service offices have begun publishing information on the 2014 farm bill programs.[16]

---

[16] USDA Farm Service Agency, *What's in the 2014 Farm Bill for Farm Service Agency Customers*, Fact Sheet, Washington, DC, March 2014, http://www.fsa.usda.gov/Internet/FSA_File/2014_farm_bill_customers.pdf. Examples for universities and state extension service include http://www.farmdoc.illinois.edu/; http://www.farmdoc.illinois.edu/webinars/index.asp; and http://www.afpc.tamu.edu/.

USDA announced in late February 2014 that the department intends to issue the regulations for PLC and ARC in the fall of 2014.[17] Once the regulations are in place, USDA indicates that producers can then update producer information and make program choices by late 2014 and early 2015. By then, producers will likely have a general indication about 2014 crop prices, and which program (PLC or ARC) will result in a higher farm payment. The decision is irrevocable for the life of the 2014 farm bill, so producers will need to consider not just 2014 prices, but also expected prices through 2018. Also, wheat farmers can enroll in the Supplemental Coverage Option policy (authorized in the 2014 farm bill) but can later opt out if they select ARC.

---

[17] Tom Vilsack, "Secretary Vilsack Addresses Commodity Classic on Farm Bill Implementation," San Antonio, Texas, February 28, 2014, http://www.usda.gov/wps/portal/usda/usdahome?contentid=2014/02/0032 xml.

Appendix. Major Farm Commodity Provisions in the Enacted 2014 Farm Bill

Table A-1. Major Farm Commodity Provisions in the 2008 Farm Bill and 2014 Farm Bill

Prior Law: 2008 Farm Bill (P.L. 110-246)	Current Law: 2014 Farm Bill (P.L. 113-79)
Program suspension	
Suspends the permanent price support authority of the Agricultural Adjustment Act of 1938 and the Agricultural Adjustment Act of 1949 for the 2008-13 crop years (covered commodities, peanuts, and sugar), and for milk through December 31, 2013. *[7 U.S.C. 8782]*	Same as prior law, except applies to 2014-2018 crop years, and through December 31, 2018 for milk. *[Sec. 1602]*
Program elimination	
Direct payments (DPs) are available to producers on farms with base acres (historical plantings) of covered commodities (wheat, corn, grain sorghum, barley, oats, upland cotton, rice, soybeans, and other oilseeds). *[7 U.S.C. 8713]* Covers 2008-2013 crop years. Direct payment rates are fixed in statute *[7 U.S.C. 7913(b)]* and do not vary based on market price. Payment amount = direct payment rate, times 85% of base acres *[7 U.S.C. 7911]*, times direct payment yield *[7 U.S.C. 7912]*. (Exception: payment acreage is 83.3% of base acres for crop years 2009-2011.) Direct payments for peanuts authorized separately. *[7 U.S.C. 8753]*	Repeals direct payments. *[Sec. 1101]*

Transition payments are made available for upland cotton for the 2014 crop year. Payment is also made for upland cotton 2015 if STAX crop insurance authorized in Title XI is not available. *[Sec. 1119]* |
| **Counter-cyclical payments (CCPs)** are available for covered commodities for 2008-2013 crop years. *[7 U.S.C. 8714]* Payment rate is difference between target price in statute (see below) and national average market price (or loan rate, if higher), minus the direct payment rate. Counter-cyclical payments for peanuts authorized separately. *[7 U.S.C. 8754(a)(1)-(3)]* | Repeals counter-cyclical payments. *[Sec. 1102]* |
| For covered commodities and peanuts, **Average Crop Revenue Election (ACRE) payments** are available to producers as an alternative to CCPs. Revenue payment based on a two-part trigger: (1) if actual state revenue is less than a guaranteed state level for the commodity, and (2) if actual farm revenue is less than a farm ACRE benchmark for the commodity. Payment amount equals the product of (1) the lesser of (a) the ACRE program guarantee minus actual state revenue or (b) 25% of the ACRE program guarantee, times (2) 83.3% (for crop years 2009-2011) or 85% (2012-2013) of the acreage planted of the covered commodity (not to exceed base acres of the commodity), times (3) the 5-year Olympic average farm yield divided by the 5-year Olympic average state yield (Olympic average drops lowest and highest year). For producers who participate in ACRE, loan rates under the marketing assistance loan program are reduced 30% and direct payments are reduced by 20%. *[7 U.S.C. 8715]* | Repeals Average Crop Revenue Election (ACRE) program. *[Sec. 1103]* |

Prior Law: 2008 Farm Bill (P.L. 110-246)	Current Law: 2014 Farm Bill (P.L. 113-79)
Definitions:	
No comparable definition. However, the terms "actual state revenue" and "actual farm revenue" were used in the Average Crop Revenue Election (ACRE) program (see below).	**Actual crop revenue:** the amount determined by the Secretary of Agriculture (the Secretary) under the Agriculture Risk Coverage program in Section 1117(b) for each covered commodity for a crop year. *[Sec. 1111(1)]*
No comparable definition.	**Agriculture Risk Coverage (ARC):** "shallow loss" revenue coverage provided under Section 1117. *[Sec. 1111(2)]*
No comparable definition. However, the term "ACRE program guarantee" was used in the Average Crop Revenue Election (ACRE) program (see below).	**Agriculture Risk Coverage (ARC) Guarantee:** the amount determined under the Agriculture Risk Coverage program in Section 1117(c). *[Sec. 1111(3)]*
Base acres: The number of base acres of a covered commodity on a farm as established under the 2002 farm bill *[7 U.S.C. 7911]*, subject to adjustments for pulse crops, other oilseeds, and conservation reserve contracts. *[7 U.S.C. 8711]* Same for peanuts. *[7 U.S.C. 7952, 7 U.S.C. 8752]*	**Base acres:** Individual crop-specific base acreages (except cotton) are retained, as in effect on September 30, 2013, subject to any adjustments in Sec. 1112, including conservation reserve contracts and inclusion of additional oilseeds designated by the Secretary. The term "base acres" includes any generic base acres planted to a covered commodity as determined in Section 1114(b). *[Sec. 1111(4)]*
No comparable definition.	**County coverage:** type of coverage under the Agriculture Risk Coverage (ARC) program under Section 1115(b)(1) to be obtained by the producer at the county level. *[Sec. 1111(5)]*
Covered commodities: wheat, corn, grain sorghum, barley, oats, upland cotton, long grain rice, medium grain rice, pulse crops (dry peas, lentils, small chickpeas, and large chickpeas), soybeans, and other oilseeds (sunflower seed, rapeseed, canola, safflower, flaxseed, mustard seed, crambe, sesame seed). *[7 U.S.C. 7901(4,9)* Peanuts were not defined as a "covered commodity" but treated as such under farm programs.	Similar to previous law except upland cotton is removed and peanuts are added to the list of covered commodities. Also, acreage of wheat, oats, and barley used for haying and grazing is eligible. *[Sec. 1111(6)]*
Effective price: the higher of (1) the national season average market price or (2) national average loan rate plus the direct payment rate. *[7 U.S.C. 7914(b)]* Same provision for peanuts. *[7 U.S.C. 7954(b)]*	**Effective price:** for a covered commodity for a crop year, the price calculated by the Secretary under Section 1116(b) to determine whether Price Loss Coverage (PLC) payments are required for that crop year. *[Sec. 1111(7)]* The effective price is the higher of (1) the national average market price received by producers during the 12-month marketing year for the covered commodity, as determined by the Secretary; or (2) the national average loan rate for a marketing assistance loan. *[Sec. 1116(b)]* The effective price for barley is the all-barley price. *[Sec. 1116(f)]*.

Prior Law: 2008 Farm Bill (P.L. 110-246)	Current Law: 2014 Farm Bill (P.L. 113-79)
Extra long staple cotton: cotton that (A) is produced from pure strain varieties of the Barbadense species or any hybrid of the species, or other similar types of extra long staple cotton having characteristics needed for various end uses for which U.S. upland cotton is not suitable, and grown in irrigated or other designated U.S. cotton-growing regions; and (B) is ginned on a roller-type gin or, other authorized gin for experimental purposes. *[7 U.S.C. 7901(7)]*	**Extra long staple cotton:** same as prior law. *[Sec. 1111(8)]*
No comparable definition.	**Generic base acres:** the number of base acres for cotton in effect under Section 1001 of the Food, Conservation, and Energy Act of 2008 (7 U.S.C. 8702), as adjusted pursuant to Section 1101 of such Act (7 U.S.C. 8711), as in effect on September 30, 2013, subject to any adjustment or reduction under Section 1112 of this act. *[Sec. 1111(9)]* Generic base is eligible for payments if a covered crop is planted on the farm (see payment discussion below).
No comparable definition. However, note that under the Supplemental Revenue Assistance (SURE) program *[7 U.S.C. 1531]*, payments compensated producers for a portion of losses at the farm level that was not eligible for an indemnity payment under a crop insurance policy.	**Individual coverage:** type of coverage under the Agriculture Risk Coverage (ARC) program in Section 1115(b)(1) to be obtained by the producer at the farm level. *[Sec. 1111(10)]*
Medium grain rice: includes short grain rice. *[7 U.S.C. 8702(9)]*	**Medium grain rice:** includes short grain rice and temperate japonica rice. *[Sec. 1111(11)]*
Other oilseed: a crop of sunflower seed, rapeseed, canola, safflower, flaxseed, mustard seed, crambe, sesame seed, or, if designated by the Secretary, another oilseed. *[7 U.S.C. 7901(9)]*	**Other oilseed:** same as prior law. *[Sec. 1111(12)]*
Payment acres: 85% of the base acres of a covered commodity on a farm on which direct payments and counter-cyclical payments are made. Exception: 83.3% of base acres for direct payments only for crop years 2009-2011. *[7 U.S.C. 7901(10)]*	**Payment acres:** the number of acres determined for a farm under Section 1114 for Price Loss Coverage payments and Agriculture Risk Coverage payments. *[Sec. 1111(13)]*
Payment yield: the yield for counter-cyclical program payments under the 2002 farm bill, generally based on 1998-2001 data. *[7 U.S.C. 7901(10)]*	**Payment yield:** the yield used to make counter-cyclical payments under the 2008 farm bill as in effect on September 30, 2013; or the yield established under Section 1113 (Price Loss Coverage program) of this act. *[Sec. 1111(14)]*
No comparable definition.	**Price Loss Coverage:** price-based coverage provided under Section 1116. *[Sec. 1111(15)]*
Producer: generally, an owner, operator, landlord, tenant, or sharecropper that shares in the risk of producing a crop and is entitled to share in the crop available for marketing from the farm, or would have shared had the crop been produced. For a grower of hybrid seed, the existence of a hybrid seed contract and other program rules shall not adversely affect the ability to receive a payment. *[7U.S.C. 7901(12)]*	**Producer:** same as prior law. *[Sec. 1111(16)]*
Pulse crop: dry peas, lentils, small chickpeas, and large chickpeas. *[7U.S.C. 8702(14)]*	**Pulse crop:** same as prior law. *[Sec. 1111(17)]*

Prior Law: 2008 Farm Bill (P.L. 110-246)	Current Law: 2014 Farm Bill (P.L. 113-79)
Target prices for 2013:	**Reference prices:**
Wheat, bu., $4.17	Wheat, bu., $5.50
Corn, bu., $2.63	Corn, bu., $3.70
Grain sorghum, bu., $2.63	Grain sorghum, bu., $3.95
Barley, bu., $2.63	Barley, bu., $4.95
Oats, bu., $1.79	Oats, bu., $2.40
Upland cotton, lb., $0.7125	Upland cotton, none (covered by STAX program Title XI)
Long grain rice, cwt., $10.50	Long grain rice, cwt., $14.00
Medium grain rice, cwt., $10.50	Medium grain rice, cwt., $14.00
Soybeans, bu., $6.00	—(temperate japonica rice, reference price for PLC is increased an additional 15% to $16.10 per cwt.) *[Sec. 1116(g)]*
Other oilseeds, cwt., $12.68	Soybeans, bu., $8.40
Dry peas, cwt., $8.32	Other oilseeds, cwt., $20.15
Lentils, cwt., $12.81	Dry peas, cwt., $11.00
Small chickpeas, cwt., $10.36	Lentils, cwt., $19.97
Large chickpeas, cwt., $12.81	Small chickpeas, cwt., $19.04
Peanuts, ton, $495	Large chickpeas, cwt., $21.54
[7U.S.C. 8714(c)(3)]	Peanuts, ton, $535
	[Sec. 1111(18)]
Secretary: the Secretary of Agriculture. *[7 U.S.C. 7901(13)]*	**Secretary:** same as prior law. *[Sec. 1111(19)]*
State: each of the U.S. States, the District of Columbia, the Commonwealth of Puerto Rico, or U.S. territory/possession. *[7 U.S.C. 7901(14)]*	**State:** same as prior law. *[Sec. 1111(20)]*
No comparable definition.	**Temperate japonica rice:** rice that is grown in high altitudes or temperate regions of high latitudes with cooler climate conditions, in the Western United States, as determined by the Secretary. *[Sec. 1111(21)]*
Transitional yield: as defined in section 502(b) of the Federal Crop Insurance Act, the	**Transitional yield:** same as prior law. *[Sec. 1111(22)]*

Prior Law: 2008 Farm Bill (P.L. 110-246)	Current Law: 2014 Farm Bill (P.L. 113-79)
maximum average production per acre or equivalent measure that is assigned to acreage for a crop year by the Federal Crop Insurance Corporation whenever the producer fails to: (A) certify that acceptable documentation of production and acreage for the crop year is in the possession of the producer; or (B) present the acceptable documentation. *[7 U.S.C. 1502(b)]*	
United States: when used in a geographical sense, all of the States. *[7 U.S.C. 7901(16)]*	**United States:** same as prior law. *[Sec. 1111(23)]*
United States premium factor: the percentage by which the difference in the United States loan schedule premiums for Strict Middling (SM) 1 1/8-inch upland cotton and for Middling (M) 1 3/32-inch upland cotton exceeds the difference in the applicable premiums for comparable international qualities. *[7 U.S.C. 8702(18)]*	**United States premium factor:** same as prior law. *[Sec. 1111(24)]*
Base, Payment Yields, and Payment Acres	
Base acres: Unlike 2002 farm bill, no provision in 2008 farm bill for reallocating base acres among crops to reflect recent plantings.	**Base acre retention or one-time reallocation:** Farm owners have a one-time opportunity to either: (1) retain base acres (including any generic base acres), or (2) reallocate base acres (other than any generic base acres) among covered crops to reflect recent historical planting shares equal to the 4-year average (2009-2012) planted area for each covered crop (including prevented plant acreage). Additional provisions: all four years are included in the average (even if zero); with multiple plantings per years, only one crop may be used in the calculation; reallocation of base acres among covered commodities may not result in an increase in total base acres (including generic base). *[Sec. 1112(a)]*
Accommodations for double cropping and CRP acres. Base cannot exceed total cropland. *[7 U.S.C. 7911]*	Similar to prior law, including provision to increase/decrease base when land leaves/enters conservation programs. *[Sec. 1112(b)]* However, base is reduced if the sum of the base acres for the farm (including any new oilseed acreage and generic base acres) plus any acreage in CRP or the Wetlands Reserve Program (or other federal conservation program that makes payments in exchange for not producing a crop) exceeds the actual cropland acreage on the farm. *[Sec. 1112(c)]* Base is reduced proportionately when acreage has been subdivided and developed for multiple residential units or other non-farming uses. *[Sec. 1112(d)]*
Payment yields: Established direct payment yields for designated oilseeds, camelina, or pulse crops using 1998-2001 farm yields, adjusted back to the national average from 1981-85. For counter-cyclical payments, yield could be updated to reflect yields during 1998-2001 using specific formulas. *[7 U.S.C. 7912]*	**Payment yields:** For Price Loss Coverage payments under Section 1116, the Secretary shall establish a yield for each farm for any designated oilseed for which a payment yield was not established under Section 1102 of the 2008 farm bill. If the yield per planted acre for a crop of a designated oilseed for a farm for any of the 1998 through 2001 crop years was less than 75% of the county yield for that designated oilseed, the Secretary shall assign a yield for that crop year equal to 75% of the county yield. *[Sec. 1113(a-c)]*
No comparable provision.	**Updating payment yields for Price Loss Coverage (PLC):** The owner of a farm has a 1-time opportunity to update, on a covered commodity-by-covered-commodity basis, the payment yield used in calculating PLC payments for each covered commodity. The PLC payment yield is equal to 90% of the average of the yield per planted acre for the crop of the

Prior Law: 2008 Farm Bill (P.L. 110-246)	Current Law: 2014 Farm Bill (P.L. 113-79)
	covered commodity for the 2008 through 2012 crop years, excluding any crop year in which the acreage planted to the crop of the covered commodity was zero. If the yield for any of the 2008 through 2012 crop years was less than 75% of the average county yield during that period, a "plug" yield for that crop year is equal to 75% of the county average for 2008 through 2012. *[Sec. 1113(d)]*
Payment acres: 85% of the base acres of a covered commodity on a farm on which direct payments and counter-cyclical payments are made. Exception: 83.3% of base acres for direct payments only for crop years 2009-2011. *[7 U.S.C. 7901(10)]*	**Payment acres:** For Price Loss Coverage (PLC) and the Agriculture Risk Coverage (ARC) when county coverage has been selected, the number of payment acres for each covered commodity on a farm is equal to 85% of the base acres for the covered commodity. In the case of ARC when individual coverage has been selected, the payment acreage is equal to 65% of the base acres for all of the covered commodities on the farm. *[Sec. 1114(a)]* Payment acres may not include any crop subsequently planted during the same crop year on the same land for which the first crop is eligible for PLC payments or ARC payments, unless the crop was approved for double cropping in the county, as determined by the Secretary. *[Sec. 1114(c)]*
	Generic base is eligible for payments if a covered crop is planted on the farm. Specifically, for each crop year, generic base acres are attributed (i.e. temporarily designated as) base acres to a particular covered commodity base in proportion to that crop's share of total plantings of all covered commodities in that year. The amount of generic base attributed for a particular year cannot exceed the acreage planted to covered crops in that year. In other words, if the total number of acres planted to all covered commodities on the farm does not exceed the generic base acres on the farm, only the amount of acreage actually planted to a covered commodity is attributed to that covered commodity for payment purposes. *[Sec. 1114(b)]*
Farms with limited base acres: A producer on a farm may not receive direct payments, counter-cyclical payments, or average crop revenue election payments (see below) if the sum of the base acres of the farm is 10 acres or less (provision was suspended for the 2008 crop year only). *[7 U.S.C. 8711(d)]*	**Minimal payment acres:** A producer on a farm may not receive Price Loss Coverage payments or Agriculture Risk Coverage payments (see below) if the sum of the base acres on the farm is 10 acres or less, except for socially disadvantaged farmers/ranchers or limited resource farmers/ranchers, *[Sec. 1114(d)]*
Planting flexibility: Any crop may be planted on base acres to receive program benefits, except fruits, vegetables (other than mung beans and pulse crops), and wild rice. Exceptions provide for farms and producers with a history of double-cropping or history of growing fruits and vegetables; in this case, direct and counter-cyclical payments are reduced acre-for-acre for the year. *[7 U.S.C. 8717, 7 U.S.C. 8756]* A pilot program beginning in 2009 in seven Midwestern states allowed planting of fruits and vegetables for processing on base acres. Base acres were temporarily reduced for the year, but restored for the next crop year and "considered planted" for any future base calculations. *[7 U.S.C. 8717(d)]*	**Effect of planting fruits and vegetables:** Any crop may be planted without effect on base acres. However, payment acres on a farm are reduced in any crop year in which fruits, vegetables (other than mung beans and pulse crops), or wild rice have been planted on base acres. The reduction to payment acres is one-for-one for each acre planted to these crops in excess of 15% of base acres for either the Price Loss Coverage or county coverage under the Agriculture Risk Coverage (ARC) program, and in excess of 35% of base acres for ARC individual coverage. *[Sec. 1114(e)]*

Prior Law: 2008 Farm Bill (P.L. 110-246)	Current Law: 2014 Farm Bill (P.L. 113-79)
Producer election: Producers may choose between the Counter-cyclical payments (CCPs) and Average Crop Revenue Election (ACRE) payments. If CCP was selected, direct payments were reduced 20% and marketing loan rates are reduced by 30%. Producers can enter ACRE for any year of the 2009-2012 crop years, but cannot return to the traditional counter-cyclical program. *[7 U.S.C. 8715(a)(1)]* If all of the producers on a farm fail to make an election or make different elections, all of the producers on the farm are deemed to have made the election to receive counter-cyclical payments. *[7 U.S.C. 8715(a)(5)]* After the 2008 farm bill was extended by 1 year by the American Taxpayer Relief Act of 2012, producers were offered the same choice for the 2013 crop year.	**Producer election:** For the 2014 through 2018 crop years, all of the producers on a farm shall make a one-time, irrevocable election to obtain either Price Loss Coverage on a covered commodity-by-covered-commodity basis; or Agriculture Risk Coverage. *[Sec. 1115(a)]* For ARC, producers must unanimously select whether to receive ARC payments based on: (1) county coverage applicable on a covered-commodity-by-covered-commodity basis; or (2) individual coverage applicable to all of the covered commodities on the farm. *[Sec. 1115(b)]* If all the producers on a farm fail to make a unanimous election for the 2014 crop year, no program payments are made to the farm for the 2014 crop year and the producers on the farm shall be deemed to have elected PLC for all covered commodities on the farm for the 2015 through 2018 crop years. *[Sec. 1115(c)]* If all the producers on a farm select ARC county coverage for a covered commodity, the Secretary cannot make PLC payments to the producers on the farm with respect to that covered commodity. *[Sec. 1115(d)]* If all the producers on a farm select individual coverage, payment calculations include the producer's share of all farms in the same state in which the producer has an interest and for which individual coverage has been selected. *[Sec. 1115(e)]* Producers on a farm cannot reconstitute the farm to void or change a program election. *[Sec. 1115(f)]*
Price-Based Payments	
Counter-cyclical payments (CCPs) are available for covered commodities for 2008-2013 crop years when the farm price is below the effective price (target price in statute minus the direct payment rate). *[7 U.S.C. 8714]*	Repeals counter-cyclical payments. *[Sec. 1102]*
	Establishes similar program called **Price Loss Coverage (PLC)** for crop years 2014-18. PLC payments are made on a farm on a covered commodity-by-covered-commodity basis if the effective price (the higher of the national average 12-month market price or the loan rate) is less than the reference price.
Payment rate is the difference between target price in statute and the national 12-month average farm price (or loan rate, if higher), minus the direct payment rate. Counter-cyclical payments for peanuts authorized separately. *[7 U.S.C. 8754(a)(1)-(3)]*	**Payment rate** is the difference between the reference price in statute and the national 12-month average market price (or loan rate, if higher). *[Sec. 1116]*
Payment amount = Payment rate times 85% of base acres times counter-cyclical program yield for the farm (generally based on 1998-2001 data). *[7 U.S.C. 7912]*	**Payment amount =** Payment rate times 85% of base acres for each covered commodity (including attributed generic base) times existing counter-cyclical program yield (or updated yields equal to 90% of 2008-2012 average yield per planted acre). *[Sec. 1116(d)]* (See Sec. 1113 for more on payment yields.)
	Payments shall be made beginning October 1, or as soon as practicable thereafter, after the end of the applicable marketing year for the covered commodity. *[Sec. 1116(e)]*

Prior Law: 2008 Farm Bill (P.L. 110-246)	Current Law: 2014 Farm Bill (P.L. 113-79)
Revenue-Based Payments	
For covered commodities and peanuts, **Average Crop Revenue Election (ACRE) payments** are available to producers as an alternative to CCPs. *[7 U.S.C. 8715]*	Repeals Average Crop Revenue Election (ACRE) program. *[Sec. 1103]*
	Establishes similar program called **Agriculture Risk Coverage (ARC)** as an alternative to PLC payments for crop years 2014-18 on a commodity-by-commodity basis for each farm (except when producers select individual (farm) ARC, then PLC is not an option for any commodity). Producers select either county coverage or individual farm coverage. *[Sec. 1117]*
Revenue payment based on a two-part trigger: (1) if the actual state revenue is less than a guaranteed state level for the commodity, and (2) if actual farm revenue is less than a farm ACRE benchmark for the commodity. *[7 U.S.C. 8715(b)(2)]*	**Revenue payment based on single trigger:** if the actual crop revenue is less than the agriculture risk coverage guarantee. *[Sec. 1117(a)]*
Actual state revenue per acre = actual state yield, times the national average market price. Actual state yield is the actual quantity produced in the state during the crop year, divided by planted acres. *[7 U.S.C. 8715(c)]* National average market price is the greater of the national average price received during the 12-month marketing year, or the marketing loan rate after being reduced by 30%. *[7 U.S.C. 8715(a)(1)]*	In the case of county coverage, **actual crop revenue** per acre = the actual average county yield per planted acre for the covered commodity times the higher of (i) the national average market price received by producers during the 12-month marketing year, or (ii) the national average loan rate. *[Sec. 1117(b)(1)]*
Actual farm revenue per acre = actual farm yield, times the greater of the national average price received during the 12-month marketing year or the marketing loan rate after being reduced by 30%. *[7 U.S.C. 8715(e)]*	In the case of individual (farm) coverage, **actual crop revenue** per acre is the producer's share of the aggregated revenue amount for all covered commodities planted on all farms for which individual coverage has been selected. Actual crop revenue is equal to the sum of covered commodity revenue (total production of each covered commodity on such farms, times the higher of (i) the national average market price received by producers during the 12-month marketing year, or (ii) the national average loan rate) divided by the total planted acres of all covered commodities on such farms. *[Sec. 1117(b)(2)]*
ACRE program guarantee per acre = 90% times the benchmark state yield, times the ACRE program guarantee price. *[7 U.S.C. 8715(d)]*	**ARC guarantee** per acre = 86% times the benchmark revenue. *[Sec. 1117(c)(1)]*
The benchmark state yield is a 5-year Olympic average state yield. *[7 U.S.C. 8715(d)(2)(A)]* The ACRE program guarantee price is a 2-year average of the national average market price. *[7 U.S.C. 8715(d)(3)]* The ACRE program guarantee cannot change more than 10% from the previous year. *[7 U.S.C. 8715(d)(1)(B)]*	In the case of county coverage, **benchmark revenue** per acre = the average historical county yield for the most recent 5 crop years (excluding the years with the highest and lowest yields, or "Olympic average") times the national Olympic average market price received by producers during the 12-month marketing year for the most recent 5 crop years. *[Sec. 1117(c)(2)]*
Farm ACRE benchmark revenue per acre = the 5-year Olympic average farm yield, times the ACRE program guarantee price; plus the crop insurance premium per acre. *[7 U.S.C. 8715(f)]*	In the case of individual coverage, **benchmark revenue** is based on the producer's share of all covered commodities planted on all farms for which individual coverage has been selected and in which the producer has an interest. Benchmark revenue is the summation of Olympic 5-year average benchmark revenue for each covered commodity aggregated across all farms with individual coverage, adjusted to reflect current-year planted acreage. Annual revenue for each covered commodity is equal to the farm yield per planted acre for the covered commodity, times the national average market price received by producers during the 12-month marketing year. *[Sec. 1117(c)(3)]*

Prior Law: 2008 Farm Bill (P.L. 110-246)	Current Law: 2014 Farm Bill (P.L. 113-79)
	If the yield per planted acre for the covered commodity or historical county yield per planted acre for the covered commodity for any of the 5 most recent crop years is less than 70% of the transitional yield, the amounts used for any of those years is equal to 70% of the transitional yield. [*Sec. 1117(c)(4)*]
	The reference price is used if the national average farm price for any of the 5 most recent crop years is lower than the reference price. [*Sec. 1117(c)(5)*]
If more than 25% of a state's acreage is irrigated and 25% is non-irrigated, separate guarantees shall apply. [*7 U.S.C. 8715(d)(4)*]	To the maximum extent practicable, calculate separate ARC guarantees for irrigated and nonirrigated covered commodities. [*Sec. 1117(g)(2)*]
Payment amount = the product of (1) the lesser of (a) the ACRE program guarantee minus actual state revenue or (b) 25% of the ACRE program guarantee, times (2) 83.3% (2009-2011) or 85% (2012 and 2013) of the acreage planted of the covered commodity (not to exceed base acres of the commodity), times (3) the 5- year Olympic average farm yield divided by the 5-year Olympic average state yield. This formula multiplies a state-level payment rate per acre (up to a maximum of 25% of the guarantee level) times a percentage of planted acreage, then prorates the payment based on the farm's yield history compared to the state's yield history. [*7 U.S.C. 8715(g)*]	The **payment rate** for a covered commodity, in the case of county coverage, or a farm, in the case of individual coverage, is equal to the lesser of: (1) the amount that the ARC guarantee for the crop year exceeds the actual crop revenue for the crop year, or (2) 10% of the benchmark revenue for the crop year. [*Sec. 1117(d)*] **Payment amount** = the payment rate times the payment acres. For county coverage, payment acres = 85% of the base acres for the covered commodity. For individual coverage, payment acreage = 65% of the base acres for all of the covered commodities on the farm. [*Sec. 1117(e)*]
For producers who participate in ACRE, loan rates under the marketing assistance loan program are reduced 30% and direct payments are reduced by 20%. [*7 U.S.C. 8715(b)(3)*]	**Timing:** Beginning October 1after the end of the marketing year. [*Sec. 1117(f)*]
Timing: Beginning October 1 after the end of the marketing year.	**Additional duties of the Secretary:** Use all available information and analysis, including data mining, to check for anomalies in the determination of ARC payments; if necessary, assign an average yield for a farm or county on the basis of the yield history representative of the area. [*Sec. 1117(g)*]
No comparable provision.	

Conservation Compliance/Producer Agreement

Eligibility for direct payments, counter-cyclical payments, or average crop revenue election payments requires producers to comply with conservation, wetland, and planting flexibility requirements; use base acres for agricultural or conserving use, and not for nonagricultural commercial, industrial, or residential use; control noxious weeds and maintain sound agricultural practices. Producers must submit annual acreage reports for all cropland on the farm; for those who receive ACRE payment, annual production reports are also required. [*7 U.S.C. 8716(a)(c)*] Same provision for peanuts (except production report requirement). [*7 U.S.C. 8755(a)(c)*] Under Title II (Conservation) of the 2008 farm bill (P.L. 110-246), benefits under the marketing loan program are subject to conservation compliance for highly erodible land [*16 U.S.C. 3811(a)(1)(A)*] and for Swampbuster [*16 U.S.C. 3812(a)(1)*].	Similar to prior law, with application to the new Price Loss Coverage (PLC) and Agriculture Risk Coverage (ARC) programs, and marketing loans. [*Sec. 1118(a)*] Annual production reports are required for producers selecting individual ARC coverage. [*Sec. 1118(c)*] No penalties can be assessed against a producer for an inaccurate acreage or production report unless the Secretary determines that the producer knowingly and willfully falsified the report. [*Sec. 1118(e)*] Separately, in Title II (Conservation), a producer must be in compliance with highly erodible land conservation requirements and wetland requirements in order to receive crop insurance premium subsidies. [*Sec. 2609*]

Prior Law: 2008 Farm Bill (P.L. 110-246)	Current Law: 2014 Farm Bill (P.L. 113-79)
Sets requirements for transfer of interest in base acres. Protects interests of tenants and sharecroppers and provides for sharing of payments on a farm on an equitable basis. *[7 U.S.C. 7915(b)(d)(e)]* Same provision for peanuts. *[7 U.S.C. 7955(b)(d)(e)]*	

Transition Payments

Prior Law: 2008 Farm Bill (P.L. 110-246)	Current Law: 2014 Farm Bill (P.L. 113-79)
No comparable provision.	Transition (direct) payments are made for upland cotton for the 2014 crop year, and for 2015 if STAX insurance product (Title XI) is not available.
	Payment equals program yield (divided by the national yield of 597 pounds per acre) *times* transition assistance rate *times* payment acres. Transition rate is based on cotton price decline between June 2013 and December 2013. Payment acres in 2014 equal 60% of 2013 cotton base acres and 36.5% in 2015. *[Sec. 1119]*

Nonrecourse Marketing Loans and Other Recourse Loans

Prior Law: 2008 Farm Bill (P.L. 110-246)	Current Law: 2014 Farm Bill (P.L. 113-79)
Nonrecourse marketing loans are available for any amount of a loan commodity (see list below) produced in crop years 2008-2013. *[7 U.S.C. 8731]* Nonrecourse marketing loans for peanuts are authorized separately. *[7 U.S.C. 8757]*	Generally continues prior law to cover 2014-2018 crop years for all loan commodities (including peanuts). *[Sec. 1201]*
For peanuts, nonrecourse marketing loans available in crop years 2008-2013. May be obtained through a marketing cooperative or association approved by USDA. Storage to be provided on a non-discriminatory basis and under any additional requirements. Payment of peanut storage costs authorized for 2008-2013 crops. *[7 U.S.C. 8757(a)(4)-(7)]*	
	Loan commodities same as prior law. *[Sec. 1201]*
Loan commodities and loan rates:	For 2014-2018 crop years, loan rates same as prior law except for upland cotton. The loan rate for upland cotton is changed from $0.52 per lb. to the simple average of the adjusted prevailing world price for the two immediately preceding marketing years, but not less than $0.45 per pound or more than $0.52 per pound. *[Sec. 1202]*
Wheat, per bushel (bu.), $2.94 ($2.75 in 2008, 2009)	
Corn, bu., $1.95	
Grain sorghum, bu., $1.95	
Barley, bu., $1.85	
Oats, bu., $1.33	
Upland cotton, lb., $0.52	
Extra long staple (ELS) cotton, lb., $0.7977	
Long grain rice, hundredweight (cwt.), $6.50	
Medium grain rice, cwt., $6.50	
Soybeans, bu., $5.00	
Other oilseeds, cwt., $10.09 ($9.30 in 2008, 2009)	
Dry peas, cwt., $5.40 ($6.22 in 2008)	
Lentils, cwt., $11.28 ($11.72 in 2008)	
Small chickpeas, cwt., $7.43	
Large chickpeas, cwt., $11.28 (not applicable in 2008)	
Graded wool, lb., $1.15 ($1.00 in 2008, 2009)	

Prior Law: 2008 Farm Bill (P.L. 110-246)	Current Law: 2014 Farm Bill (P.L. 113-79)
Nongraded wool, lb., $0.40 Mohair, lb., $4.20 Honey, lb., $0.69 ($0.60 in 2008, 2009) *[7 U.S.C. 8732 (a)(b)(c)]* Peanuts, ton, $355 *[7 U.S.C. 8757(b)]* Establishes a single loan rate in each county for each kind of "other oilseeds" *[7 U.S.C. 8732(d)]*	
Term of loans: 9 months after the day the loan is made; no extensions. *[7 U.S.C. 8733]* Same term for peanuts. *[7 U.S.C. 8757(c)]*	Same as prior law. *[Sec. 1203]*
Loan repayment: Loans may be repaid at the lesser of (1) the loan rate plus interest, (2) a rate based on average market prices during the preceding 30-day period, or (3) a rate determined by USDA that will minimize forfeitures, accumulation of stocks, storage costs, market impediments, and discrepancies in benefits across states and counties. Excludes upland cotton, rice, ELS cotton, confectionery and each other kind of sunflower seed (other than oil sunflower seed). *[7 U.S.C. 8734(a)]* Provides USDA authority to temporarily, and on a short-term basis only, adjust the repayment rates in the event of a severe disruption to marketing, transportation or related infrastructure. *[7 U.S.C. 8734(h)]* Similar provisions for peanuts. *[7 U.S.C. 8757(d)]*	Same as prior law. *[Sec. 1204]*
For upland cotton, long grain rice, and medium grain rice, repayment may be at the lesser of the loan rate plus interest, or the prevailing world price for the commodity adjusted to U.S. quality and location. *[7 U.S.C. 8734(b)]*	
For ELS cotton, repayment must be at the loan rate plus interest. *[7 U.S.C. 8734(c)]*	
For confectionery and other kinds of sunflower seeds (other than oil sunflower seed), loans must be repaid at the lesser of (1) the loan rate plus interest, or (2) the repayment rate for oil sunflower seed. *[7 U.S.C. 8734(f)]*	
For 2008-2011 crop years, USDA provides cotton storage payments at the same rates as provided for the 2006 crop, but reduced by 10%. Beginning with 2012 crop year, the rates are reduced by 20%. *[7 U.S.C. 8734(g)]*	Cotton storage payments reauthorized for 2014-2018 crop years with 10% rate reduction. *[Sec. 1204]*
Loan deficiency payments (LDP) are available to producers who agree to forego marketing loans. LDP computed by multiplying the payment rate (the amount that the loan rate exceeds the rate at which a marketing loan may be repaid) for the commodity times the quantity of the commodity produced. Loan deficiency payments available for unshorn pelts or hay and silage, even though they are not eligible for marketing loans. ELS cotton is not eligible. Payment rates determined using the rate in effect as of the date that producers request payment (producers do not need to lose beneficial interest). *[7 U.S.C. 8735]* Same provision	For 2014-2018 crop years, same as prior law. *[Sec. 1205]*

Prior Law: 2008 Farm Bill (P.L. 110-246)	Current Law: 2014 Farm Bill (P.L. 113-79)
for peanuts. *[7 U.S.C. 8757(e)]*	
Payments in lieu of LDP for grazed acreage of wheat, barley, oats, or triticale. *[7 U.S.C. 8736]*	For 2014-2018 crop years, same as prior law, except yield can be based on ARC or PLC if applicable or as determined by the Secretary. *[Sec. 1206]*
Special marketing loan provisions for upland cotton impose a special import quota on upland cotton through July 31, 2013, when price of U.S. cotton, delivered to a definable and significant international market, exceeds the prevailing world market price for 4 weeks. *[7 U.S.C. 8737(a)]* Limited global import quota is imposed on upland cotton when U.S. prices average 130% of the previous 3-year average of U.S. prices *[7 U.S.C. 8737(b)]*	Provisions extended without an expiration date beginning August 1, 2014. *[Sec. 1207]*
Economic adjustment assistance to users of upland cotton provides assistance to domestic users of upland cotton for uses of all cotton regardless of origin to acquire, construct, install, modernize, develop, convert, or expand land, plant, buildings, equipment, facilities, or machinery. Rate was 4¢/lb. between August 1, 2008, and July 31, 2012, and declined to 3¢/lb. effective beginning August 1, 2012. *[7 U.S.C. 8737(c)]*	Same as prior law except assistance begins August 1, 2013, at the 3¢/lb. rate. *[Sec. 1207]*
Special competitiveness program for ELS cotton provides payments to domestic users and exporters whenever the world market price for the lowest priced ELS cotton is below the prevailing U.S. price for a competing growth of ELS cotton for a 4-week period; and the lowest priced competing growth of ELS cotton is less than 134% of the loan rate for ELS cotton. Effective through July 31, 2013. *[7 U.S.C. 8738]*	Same as prior law during the period beginning on the date of enactment of this act through July 31, 2019. *[Sec. 1208]*
Recourse loans for high moisture feed grains and seed cotton are available for farms that normally harvest corn or sorghum in a high moisture condition at rates set by the USDA. For recourse loans for seed cotton, repayment is at loan rate plus interest. *[7 U.S.C. 8739]*	For 2014-2018 crop years, same as prior law except payment yield for feed grains is the lower of the PLC yield or actual yield. *[Sec. 1209]*
Adjustments of loan rates are authorized for any commodity (other than cotton) based on differences in grade, type, quality, location, and other factors. Allows county loan rates as low as 95% of the U.S. average, if it does not increase outlays; prohibits adjustments that would increase the national average loan rate. For cotton, loan rates may be adjusted for differences in quality factors. *[7 U.S.C. 8740]; [7 U.S.C. 8758]* for peanuts.	Same as prior law except removes certain mandatory provisions to quality adjustments. *[Sec. 1210]*
Payment Limitations	
Establishes the maximum amount of payments per year to a person or legal entity for the sum of all covered commodities, except peanuts. Peanuts have a separate but equal payment limitation.	Establishes a per-year limit on commodity program payments: Price Loss Coverage (PLC) payment, Agriculture Risk Coverage (ARC) payments, marketing loan gains and LDPs.
—Direct payments: $40,000	—PLC, ARC, marketing loan gains and loan deficiency payments, for the sum of all covered commodities except peanuts: $125,000. *[Sec. 1603(b)]*
—Direct payments under ACRE: $40,000 minus the reduction required for an ACRE participant.	—PLC, ARC, marketing loan gains and loan deficiency payments, for peanuts: $125,000. *[Sec. 1603(c)]*
—Counter-cyclical payments: $65,000	—cotton transition payments (2014 and 2015): $40,000. *[Sec. 1119(e)]*

CRS-28

Prior Law: 2008 Farm Bill (P.L. 110-246)	Current Law: 2014 Farm Bill (P.L. 113-79)
—ACRE payments: $65,000 plus the reduction in the limit from the direct payment limit. —Marketing loan gains/LDP: no limit. *[7 U.S.C. 1308 (a)-(d)]*	
Payments are attributed to a person by accounting for the direct and indirect ownership in any legal entity. Payments made directly to a person are combined with the person's pro rata share of payments from a legal entity. Payments to a legal entity cannot exceed the limits above, and are attributed to persons. Attribution of payments to legal entities is traced to four levels of ownership. If a payment has not been allocated to an individual after four levels of ownership, the payment to the first-level entity is reduced on a pro-rata basis. *[7 U.S.C. 1308 (e)-(h)]*	Continues other payment limit provisions such as direct attribution. Addresses active personal management (see below).
To be eligible for payments, persons must be "actively engaged" in farming. Actively engaged, in general, is defined as making a significant contribution of (i) capital, equipment or land, and (ii) personal labor or active personal management. Also, profits are to be commensurate with the level of contributions, and contributions must be at risk. Legal entities can be actively engaged if members collectively contribute personal labor or active personal management. Special classes allow landowners to be considered actively engaged if they receive income based on the farm's operating results, without providing labor or management. Spouses are considered actively engaged if the other spouse meets the qualification, allowing payment limits to be doubled. *[7 U.S.C. 1308-1]*	Does not change the existing statute regarding requirements to be actively engaged in farming; that is, it continues to allow active personal management. However, it instructs the Secretary of Agriculture to write new regulations that define "significant contribution of active personal management" (to more clearly and objectively implement 7 U.S.C 1308-1(b)(2), recognizing past difficulties). Specifically allows for different limits for varying types of farming operations, based on considerations of size, nature, and management requirements of different farming types, changes in the nature of active personal management due to advancements in farming practices, and the impact of this regulation on the long-term viability of farming operations. Regulations shall not apply to entities made solely of family members. Conferees intend for regional differences and a range of activities performed to be considered. Regulations are to be promulgated within six months of enactment, and may apply beginning with the 2015 crop year. *[Sec. 1604]*

Adjusted Gross Income (AGI) Limitation

Prohibits farm commodity program benefits to an individual or entity if adjusted gross income exceeds certain thresholds. For this purpose, AGI is divided into two parts: farm AGI and non-farm AGI. Uses a 3-year average when comparing to the limit.	Eliminates the distinction between non-farm AGI and farm AGI, and establishes a limit on total AGI. Uses a 3-year average when comparing to the limit. Repeals expiration date of applicability.
—$500,000 limit on non-farm AGI to qualify for and receive any farm commodity program benefits, Milk Income Loss Contract (MILC) program, noninsured crop assistance (NAP), or disaster payments.	—$900,000 limit on total AGI to qualify for and receive PLC or ARC payments, marketing loan gains or loan deficiency payments, supplemental agricultural disaster assistance, noninsured crop assistance, and (beginning in FY2015) conservation program payments. *[Sec. 1605]*
—$750,000 limit on farm AGI to qualify for and receive direct payments, but counter-cyclical, ACRE and marketing loan benefits may continue if farm AGI exceeds $750,000. *[7 U.S.C. 1308-3a(b)(1)]*	
For FY2012 only, a separate, additional $1 million AGI limit applies to direct payments *[P.L. 112-55, Sec. 745]*	
For conservation programs, $1 million limit on non-farm AGI, unless more than 66.66% of AGI	

Prior Law: 2008 Farm Bill (P.L. 110-246)	Current Law: 2014 Farm Bill (P.L. 113-79)
is farm AGI. Provides USDA discretion to waive the limit for "environmentally sensitive land of special significance." *[7 U.S.C. 1308-3a(b)(2)]*	
Supplemental Agricultural Disaster Assistance	
Beginning in 2008, five new disaster programs were authorized and funded for disasters occurring on or before 9/30/11. *[7 U.S.C. 1531]* Program funding derived from a transfer of 3.08% of annual customs receipts to the newly created Agricultural Disaster Relief Trust Fund. *[19 U.S.C. 2497(a)]* Under P.L. 112-240, all but SURE (below) were reauthorized (but not funded) for FY2012 and FY2013. To be eligible, a producer must purchase crop insurance or coverage under the Noninsured Crop Disaster Assistance Program (NAP). No funding caps except for ELAP. The five programs:	All programs (except SURE) are reauthorized permanently (retroactive to FY2012) with mandatory funding from the Commodity Credit Corporation. Producers are not required to purchase crop insurance or NAP coverage. *[Sec. 1501]*
(1) Supplemental Revenue Assistance (SURE) Payments for crops (not just farm program crops); compensated producers for a portion of losses that are not eligible for an indemnity payment under a crop insurance policy; *[7 U.S.C. 1531(b)]*	No comparable provision.
(2) Livestock Indemnity Program (LIP) compensated ranchers at a rate of 75% of market value for livestock death losses in excess of the normal mortality due to adverse weather, including hurricanes, floods, blizzards, disease, wildfires, extreme heat, and extreme cold. *[7 U.S.C. 1531(c)]* Livestock includes: cattle (including dairy cattle), bison, poultry, sheep, swine, horses, and other livestock, as determined by the Secretary. *[7 U.S.C. 1531(a)(12)]*	Same a prior law except: (1) adds as an eligible loss attacks by animals reintroduced into the wild by the federal government or protected by federal law, including wolves and avian predators; and (2) ensures that LIP payments do not duplicate any federal compensation associated with federal quarantine and disposal. *[Sec. 1501(b)]*
(3) Livestock Forage Disaster Program (LFP) compensated eligible livestock producers for grazing losses due to either (a) qualifying drought conditions (using the drought monitor system for classifying drought), or (b) fire on public managed land. An eligible livestock producer is an owner, lessee, or contract grower that provides the pastureland or grazing land and meets other criteria. *[7 U.S.C. 1531(d)]*	Same as prior law except as noted below. *[Sec. 1501(c)]*
For drought, the monthly payment rate equals 60% of estimated feed costs; number of monthly payments:	
(a) 1 month for land located in a county with a D2 drought intensity for at least 8 consecutive weeks;	
(b) 2 months for land in a county with at least a D3 rating at any time during the normal grazing period;	Increased to 3 months of payments.
(c) 3 months if the county has a D3 rating for at least 4 weeks;	Increased to 4 months of payments.
(d) 3 months if the county has a D4 rating at any time during the normal grazing period.	Increased to 4 months of payments.

Prior Law: 2008 Farm Bill (P.L. 110-246)	Current Law: 2014 Farm Bill (P.L. 113-79)
No comparable provision.	5 months of payments if the county has a D4 rating for at least 4 weeks.
Grazing losses must be on land that is native or improved pastureland, or is planted to a crop. Livestock in feedlots are not covered.	
For fire on public land, the monthly payment rate equaled to 50% of estimated feed costs and covered the period the federal agency excludes the producer from using the managed rangeland for grazing.	
(4) Emergency Assistance for Livestock, Honeybees, and Farm-Raised Catfish (ELAP) provided up to $50 million annually to compensate producers for disaster losses due to disease, adverse weather, or other conditions, such as blizzards, wildfires, and feed shortages not covered under SURE, LIP, or LFP. *[7 U.S.C. 1531(e)]*	Same as prior law except reduces maximum funding for ELAP to $20 million annually and adds cattle tick fever as a disease covered by the program. *[Sec. 1501(d)]*
(5) Tree Assistance Program (TAP) provided payments to eligible orchardists and nursery growers who suffered tree mortality greater than 15% due to a natural disaster, including plant disease, insect infestation, drought, fire, freeze, flood, earthquake, lightning, or other occurrence. For damage or mortality in excess of 15%, reimbursement equals 70% of the cost of replanting trees (or sufficient seedlings to reestablish a stand) and 50% of the cost of pruning/removal to salvage existing trees or prepare the land to replant trees. *[7 U.S.C. 1531(f)]*	Same as prior law except TAP payment rate for replanting is reduced from 70% to 65%. *[Sec. 1501(e)]*
Disaster program payment limit: $100,000 per person per year for first four programs combined. *[7 U.S.C. 1531(h)(2)]* TAP has a separate limit of $100,000 per year, and TAP payment acreage may not exceed 500 acres. *[7 U.S.C. 1531(f)(4)]*	Combined payment limit is $125,000 per person for LIP, LFP, and ELAP. Separate limit of $125,000 applies to TAP, and 500 acre limit continues. *[Sec. 1501(f)]*
Adjusted gross income (AGI) limit: nonfarm AGI cannot exceed $500,000. *[7 U.S.C. 1531(h)(3)]* and *[7 U.S.C. 1308-3a]*	**Adjusted gross income (AGI) limit:** total AGI cannot exceed $900,000. *[Sec. 1605(a)(2)]*

Source: CRS.

Notes: Provisions for the dairy and sugar programs, as well as administrative provisions in Title I, are summarized in CRS Report R43076, *The 2014 Farm Bill (P.L. 113-79): Summary and Side-by-Side.*

Author Contact Information

Dennis A. Shields
Specialist in Agricultural Policy
dshields@crs.loc.gov, 7-9051